LOOKING AT SCIENCE 5

A Closer Look

David Fielding

Basil Blackwell

© 1984 David Fielding

All rights reserved. No part of this publication may be reproduced, stored in a retrieval system, or transmitted in any form or by any means, electronic, mechanical, photocopying, recording or otherwise, without prior permission of Basil Blackwell Publisher Limited.

First published 1984

Published by Basil Blackwell Limited
108 Cowley Road
Oxford OX4 1JF

ISBN 0 631 91390 4 (*paperback*)
     0 631 13698 3 (*cased*)

Printed in Hong Kong

Topic symbols

 This work is about air and water.

 This work is about animal life.

 This work is about electricity and magnetism.

 This work is about light and dark.

 This work is about mechanics.

 This work is about plant life.

 This work is about weather and climate.

Look for the symbols in the other books in the series. There is more work about these things in the other books.

Contents

A word to teachers and parents 5

Part 1 Making things easy with science

Friction 6
What friction is and how it makes it hard to drag things. Ways of getting rid of friction.

Oil 8
What oil is like, and how it gets rid of friction. How we use oil in machines.

Wheels and tracks 10
The way that heavy things sink into soft ground. How large wheels and tracks help them.

Gears 12
What gears are, and how they let us use less effort to do things.

Levers 14
What a lever is. There are three kinds of lever. How levers make it easier to move things.

Pulleys 16
What a pulley is. How pulleys help us to lift things. A system with several pulleys makes lifting easier.

Making things easy with science 18
What bearings are, and why we use them in machines. More about oil – where it comes from and how we use it.

Part 2 Moving things with science

Sails 20
How sails of various kinds are designed to catch and use the wind.

Propellers 22
What propellers are, and how they can drive things along.

Aeroplanes and parachutes 24
Why aeroplanes are able to fly. How parachutes let us fall safely through the air.

Steam power 26
How we can use steam to drive machinery.

Electric power 28
How electricity can produce movement. Electric motors produce a powerful movement that we can use.

Jet and rocket engines 30
How jet and rocket engines work. Why they move things forward.

Moving things with science 32
A detailed look at petrol, jet and rocket engines.

Part 3 A scientific look at nature

A visit to a pond   34
How to visit a pond scientifically. What you might find there. How to take samples and record what you find.

Make an aquarium   36
How to build a home for water creatures, and how to raise frogs from frogspawn.

A visit to a seashore   38
How to explore a seashore scientifically. What to look for, and how to record your findings.

A wood in winter   40
How to explore woodland scientifically. What things to look out for, and how to investigate them.

A wood in summer   42
How to explore the same woodland in summer, to discover likenesses and differences.

A study of wild birds   44
How to make a bird table and nesting box, and use them to observe birds.

A scientific look at nature   46
Some important ways in which the sun and moon affect the Earth. The vastness of nature.

New words   48

Acknowledgements

Alan Beaumont 19(3), 33(2)
Brenard Press Limited 23(2)
Hewett Street Studios 17(3)
National Aeronautics and Space
   Administration (NASA) 31(3), cover
TI Raleigh Limited 13(3)
Tim More/Royal Yachting Association 21(3)
Science Museum 27(2)
Shell Photographic Library 19(2)
Topham Picture Library 25(2)
ZEFA 11(2,3)

Illustrations by Michael Stringer (colour)
and David Fielding (black and white)
Design by Indent, Reading

A word to teachers and parents

A Closer Look is the fifth book in the five-book *Looking at Science series*, which has been designed to do two things:
- It gives children a solid body of knowledge in natural and physical science.
- It begins to introduce them to the nature of scientific enquiry.

These two elements are developed side by side through the books.

Each double page covers a particular area for study. The left hand page outlines an activity to perform and the right hand page gives information connected with it.

The activities are introduced with the symbol ♥, and cover experimentation, observation and recording. A list of all the equipment needed for the experiments is given near the beginning of each spread.

Each book also contains another kind of double page, which is purely factual, spaced at regular intervals. These pages draw together the themes of the preceding pages.

These books can be worked through in order. Alternatively, they can be used as source material for topic work. Suggested topic areas are identified, with symbols, in the contents list.

A Closer Look deals with areas and activities especially suitable for children who show a particular interest in science.

Part 1 Making things easy with science
This section has as its theme the way in which people have learned to use scientific principles to make it easier to move things. It explains what friction is and how we use wheels and lubricants to overcome it. We use different sizes of wheels and gears to help vehicles to move more easily; levers enable us to move heavy weights, and pulleys let us lift them. Children will be able to see science being applied in everyday life.

Part 2 Moving things with science
The theme of this section is that of using science to transport things. It looks at sailing ships, propellers, aeroplanes, steam engines, electric motors, jet engines and rocket engines. Finally, it looks at petrol, jet and rocket engines in detail.

Part 3 A scientific look at nature
This section gives a broad outline framework for scientific studies of bird life, of woodland in both winter and summer, of a pond and of a seashore. It reminds the children of how to carry out their investigations scientifically – they have to decide what they want to examine, how to go about it and how to record their findings.

Part 1 Making things easy with science

Friction

Men are not strong, yet they can move heavy things. How can they do this?

You will need a container, a spring balance, two 1 kilogram weights, marbles, round pencils, Plasticine, card, Sellotape, scissors

Picture 1 An experiment to test the effort of moving something

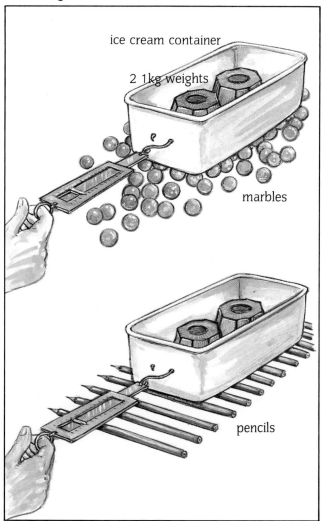

♥ *Experiment: To show the effort of moving something*

Put two 1 kilogram weights into a container. Fix a spring balance to it, and see where the pointer comes on the scale when the spring is slack. Then use the balance to pull the container along slowly. See where the pointer comes to when the spring is stretched. This stretch shows the effort you need to pull the container.

♥ *Experiment: To make the effort less*

Set out some marbles on a tray, and put the container on them. Use the spring balance to pull the container gently, and see where the pointer comes to on the scale.

Lay a track of rollers. Space them apart. Place the container on one end. Pull it along gently and read the scale on the spring balance.

Make some axles and wheels, as the picture shows. Tape them to the container. Pull the container along again, and note the reading on the scale.

♥ *Record*

1 Copy out the chart. Fill it in to show what you found out in your experiment.

The job being done	Reading on the scale
The container standing still	
The container pulled along the desk	
The container pulled on marbles	
The container pulled on rollers	
The container pulled on wheels	

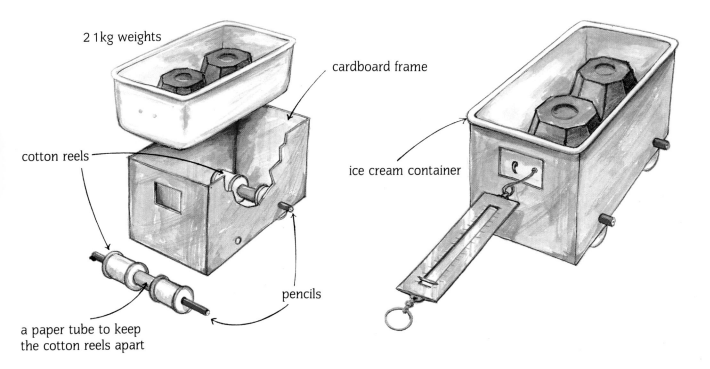

Picture 2 Arrange your experiment like this to test how friction can be reduced

2 Describe your experiments, using pictures to help.
3 Say how we can make it easier to move things. What kinds of shapes do this best?
4 Why are wheels and axles more useful than balls or rollers?

Overcoming friction

When two things rub against each other, there is a drag. This drag is caused by a force called *friction*. Friction makes it hard to pull large, heavy things. We can make friction less by using round things. We put them between the ground and the weight we are pulling. The round things roll under the weight and stop most of the drag.

Completely round things, like marbles, are good at getting rid of friction. (Though they do not get rid of it completely.) The trouble is that they roll all over the place. Rollers are better, because they will stay in a line. Early men pulled blocks of stone over rollers.

Even rollers were not a perfect help, though. Men had to keep moving rollers from the back to the front. Otherwise the stone would fall off the end.

The wheel

The wheel was an important invention. A wheel is fixed to an object, and moves along with it. Yet it rolls at the same time, and gets rid of friction.

❤ To write

1 Write down two or three ways in which friction is useful to us.
2 In which of these things is friction working against us: cycling, sleeping, walking, pulling something, skipping, writing, watching television.
3 What would happen in a world with no friction?

Oil

Modern engines have parts that slide together quickly. How do we deal with the friction between them?

> You will need silver foil, scissors, masking tape, a container, a spring balance, engine oil, weights, a jar, water, a ruler

♥ Experiment: What is oil like?

Rub a little engine oil between your finger and thumb. See how it makes a slippery covering over your skin.

Half-fill a jar with water. Pour a little oil into the water. Put the top on the jar and shake it. The oil will not mix with the water. It will float in a thick layer on top of the water.

♥ Experiment: How does oil cut down friction?

Make a track from silver foil and tape it on your desk. Turn up its edges. Pull a heavy container along the track with a spring balance. Read the scale to see where the pointer comes. Then smear engine oil on the track. Pull the container along again, and read the scale once more. See how the oil lets the container slip along more easily.

♥ Record

1 Describe your experiments, and draw them. Say what happened in each one.
2 Say what oil does to friction. Why do we use oil in machines?
3 What happens to the sea if lots of oil escapes into it?

Metal and Friction

Machines have many metal parts that rub together. Friction makes the metal hot. When metal gets hot, it grows slightly bigger. It *expands*. If the parts in a machine expand too much, they jam together and ruin the machine.

Picture 1 Testing to find out if oil cuts down friction

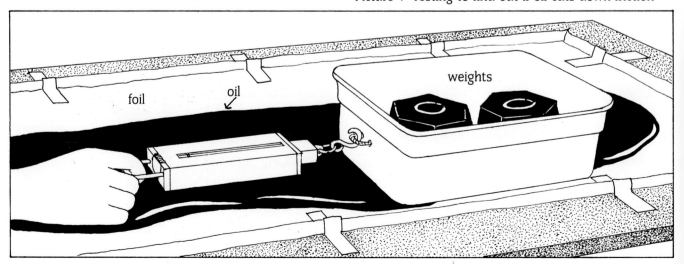

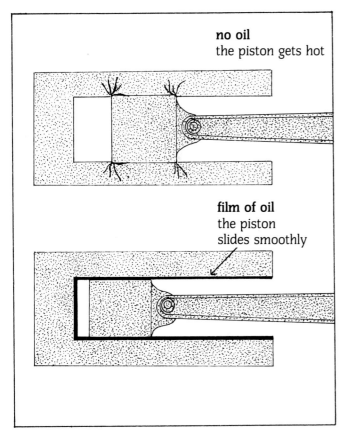

Picture 2 Oil helps engines to work better

Also, too much friction makes the parts wear away quickly. So scientists worked to get rid of friction in machines.

They learned to put oil in machines. Oil is thick and slippery. It slides between all the parts of a machine. It makes a slippery layer between the moving parts, and lets them slide past each other easily. It gets rid of most of the friction. It stops the metal from getting too hot and expanding too much. It also makes the machines last longer.

♥ To write

1 Oil _____ friction. (increases, decreases, heats, expands)
2 What would happen if an engine ran without any oil inside it?
3 Why do machines last longer when they are oiled?

Picture 3 Machines which are oiled run smoothly and quietly

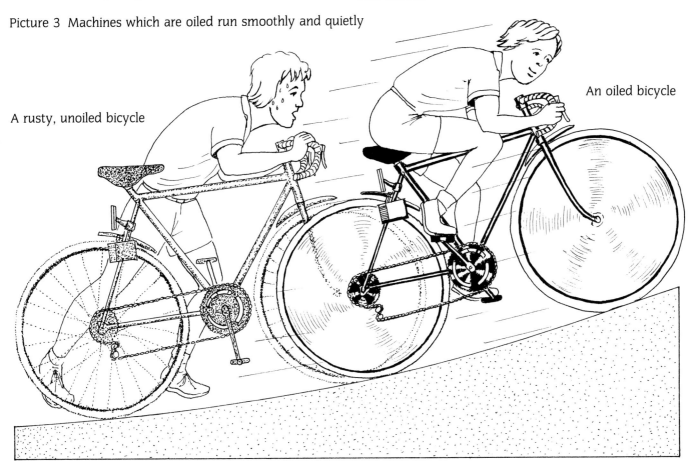

Wheels and tracks

Why are tractor wheels so big and wide? Why do tanks have tracks instead of wheels?

> You will need sawdust, four narrow cotton reels, pencils, a container, heavy weights, Sellotape, card, scissors, a ruler

♥ Experiment: Why do things sink into soft ground?

Fill a deep container with sawdust. Sellotape four pencils into a cardboard square, as the picture shows. Lay this square on the sawdust. Put a heavy container on it. Mark the wooden 'legs' to show how far they sink into the sawdust. Take the container off and lift out the legs. Measure the length of wood that sank into the sawdust.

Repeat the experiment with pencils taped together to make legs twice as wide. Measure how far they sink, and compare this measurement with your first one. Then try using narrow cotton reels in place of the pencils.

Last, lay the container on the sawdust without any legs under it. The large area of the container may not sink at all.

♥ Record

1 Make this chart to show your results.

What supported the weight	How deep it sank
Thin pieces of wood	cm
Wider pieces of wood	cm
Narrow cotton reels	cm
The whole container	cm

2 Describe what you did, and what you found. Draw the experiment.

Picture 1

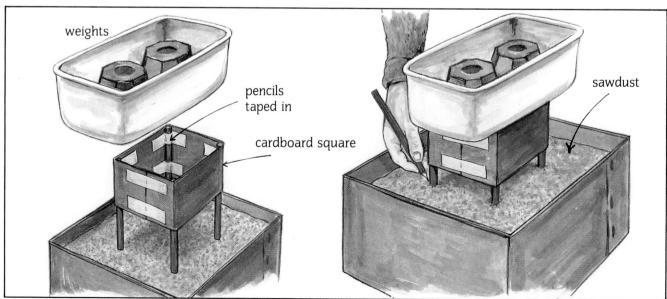

3 Are small areas good at supporting heavy things? Does it help if we use supports with a bigger area?

4 Does much of a wheel touch the ground? Do big wheels have a larger area touching the ground?

Keeping out of soft ground

Heavy machines often have to travel over soft ground. If they had ordinary wheels, they would sink into the ground. Ordinary wheels have only a small area touching the ground. You found that when a heavy weight is carried on a small area, it sinks into the ground.

When a weight is spread over a bigger area, it does not sink so far. This is why tractors have big wheels. Big wheels have bigger areas touching the ground. They spread the weight of the tractor over a bigger area. The tractor does not sink so much in soft ground.

Picture 2 Large wheels stop things sinking into soft ground

Picture 3 Why do skis stop people sinking into snow?

Tanks have tracks instead of wheels. Tracks spread the tremendous weight of tanks over larger areas still.

Moving on snow

In snowy countries, people sometimes use snowshoes. These are large, flat shapes tied to the feet. They spread a person's weight over a large area. They let people walk on thick snow without sinking into it.

♥ To write

1 How could you make it easy to ride a bicycle on deep sand or gravel?

2 A heavy weight will sink less if we carry it on a _____ area. (smaller, thinner, larger, lower)

3 Describe another way of travelling on snow.

Gears

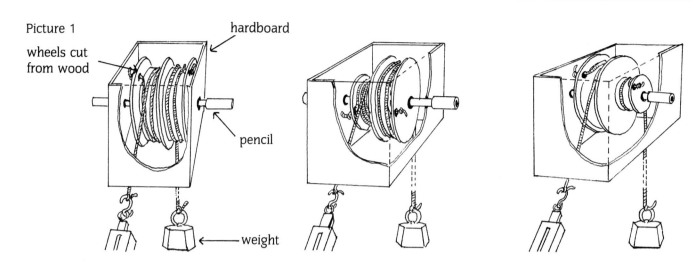

Picture 1

What are *gears*, and how do they work?

> You will need wood, hardboard, pencils, string, a ruler, a spring balance, weights

❤ Experiment: To show how gears work

Make a frame like the one in Picture 1. Fix 60 centimetres of thread to each wheel.

Put the two equal-sized wheels on the frame. Make the thread dangle from one, and tie a 200 gram weight to it. Wind the thread round the other, and tie a spring balance to the thread.

Gently pull on the spring balance until the thread is unwound. The weight will lift. Record where the pointer comes on the scale of the balance as you pull. How far did the weight lift?

Change the wheels so that a large one is with a small one. Repeat the experiment with the weight hanging from the big wheel. Then repeat it with the weight hanging from the small one. Try using different weights.

❤ Record

1 Make Chart 1.
2 How can you make a job easier by using different-sized wheels together?

Gears and machines

Gears are wheels of different sizes that are put together to make jobs easier. You

Chart 1

Wheels	Weight lifted	Reading on scale
Two wheels of equal size	200 grams	grams
A small wheel moving a big one	200 grams	grams
A big wheel moving a small one	200 grams	grams

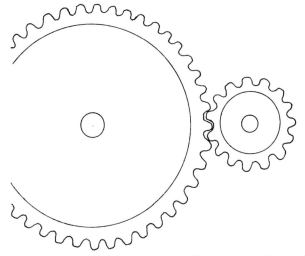

Picture 2 Gear wheels. The small one turns faster than the large one

found that using a big wheel to turn a small one made it easier to lift a weight. To lift the weight right up, you would have had to turn the big wheel for a long time. But the effort would be small.

When you used a small wheel to turn a big one, it was harder. You lifted the weight a long way up, but you needed a stronger effort.

We use gears in machines. Gears usually have teeth to join them together, or chains. Bicycles have gears joined by a chain. A large pedal gear is joined to other gears at the back wheel. You can join the pedal gear to gears of different sizes. This lets you change the amount of effort you need to use. It makes it easier for you to cycle uphill.

Gearboxes

Cars have gears in them. The driver can choose different gears for starting, climbing hills and driving on a level road. The gears are wheels of different sizes, put together into a gearbox. Many machines driven by engines have gearboxes to make their work easier.

♥ To write

1 We can make work easier by using a _____ wheel to turn a _____ one. (big, small)

2 If we use a small wheel to turn a big one, we have to use _____ effort. (great, little)

3 How does a driver choose different gears? When would you use a high gear?

Picture 3 This picture shows you where the gears are on a bike and how they work

High gear – good for flat roads. The pedal wheel is connected to a small gear wheel.

Each turn of the pedals takes a lot of effort and turns the back wheel a long way.

Low gear – good for starting off and for hills. The pedal wheel is connected to a large gear wheel.

Each turn of the pedals takes little effort but does not turn the back wheel far.

Levers

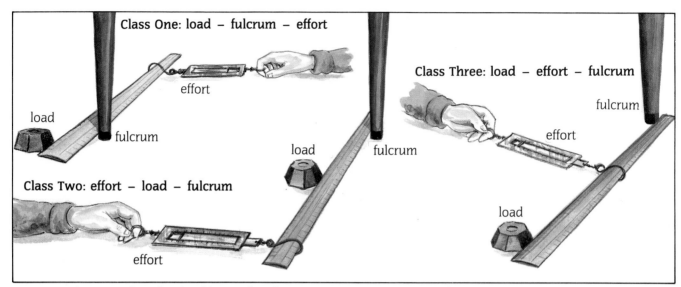

Picture 1 There are three kinds of lever. This diagram shows you how to make each type

Levers help us to move things.

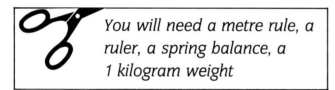

You will need a metre rule, a ruler, a spring balance, a 1 kilogram weight

♥ *Experiment: Making different kinds of lever*

Fix a spring balance to a weight and pull it along. Read the scale and record where the pointer comes. Then make the three arrangements shown in Picture 1. Pull the spring balance to move the weight. Read the scale of the balance each time.

♥ *Experiment: To show that long levers work best*

Make the arrangement called Class 1. Make sure that the desk leg touches the twenty centimetre mark on the ruler. Pull the lever with the spring balance. Pull it at the 40 centimetre, 60 centimetre, 80 centimetre and 100 centimetre marks. Read the scale of the spring balance each time, and record the reading.

♥ *Record*

1 Draw the three ways of arranging a lever. Say how you used them to move the weight. Did each lever make it easier to move the weight? Make this chart to show what you found:

	Reading on the spring balance
The weight on its own	
Moving the weight with a Class 1 lever	
Moving the weight with a Class 2 lever	
Moving the weight with a Class 3 lever	

Picture 2 A lever can help you move something which you cannot move on your own

Picture 3 Examples of the three different classes of levers

Class One: see-saw

Class Two: wheelbarrow

Class Three: arm

2 Describe how you experimented with the Class 1 lever. Make this chart to show what you found:

Where we pulled	Reading on the spring balance
On the 40 cm mark	
On the 60 cm mark	
On the 80 cm mark	
On the 100 cm mark	

3 Is it easier to pull a lever near to its anchor, or far away?

4 Why is a long lever better than a short one?

How levers help us

It always takes effort to move something. Look at the man trying to move a rock. He is using a lot of effort. But then he uses a lever, and moves the rock easily. The lever helps him in two ways. It lets him make his effort further away. This means that he can stand comfortably. It also lets him use much less effort. He has to move the end of the lever quite a long way, but he does not have to strain. Levers let us use less effort to move things.

The three classes of lever are alike in some ways. They all need an anchor, or *fulcrum*. They all need length. The longer a lever is, the better. The further from the fulcrum you can get, the easier it is to move a lever. Your second experiment showed this. All levers need a load to move.

Picture 3 shows some examples of different types of levers.

♥ To write

1 Where are the levers on a bicycle?

2 Which are levers: a desk lid, a see-saw, a table knife, a doorhandle, a spade?

3 What happens to a lever that has no fulcrum?

Pulleys

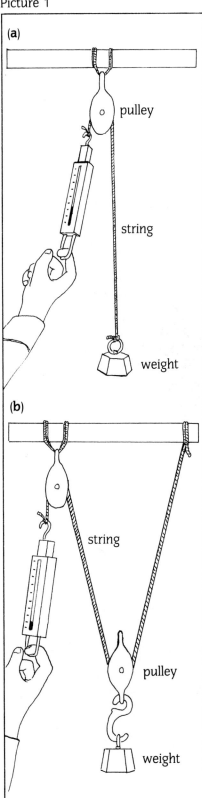

Picture 1

Simple ropes and wheels help us to lift heavy weights.

You will need two pulleys, a length of wood to hold the pulleys, a 1 kilogram weight, a spring balance, a ruler, string, scissors

♥ *Experiment: To show the effort of lifting*

Lift a 1 kilogram weight on a spring balance. The stretch of the spring shows the effort needed to lift the weight. Record where the pointer comes to on the scale.

♥ *Experiment: To show how pulleys help lifting*

Fix 60 centimetres of string to your weight and tie a spring balance to the end of it. Then arrange a *pulley* as Picture 1(a) shows. Fit your length of string over the pulley. Lift the weight by pulling down gently on the spring balance. Record where the pointer comes to on the scale of the balance.

So far, you have not changed the effort of lifting the weight. But the pulley lets you pull downwards instead of up.

Next, set up two pulleys as Picture 1(b) shows. Again, slowly lift the weight by pulling downwards. Record where the pointer comes to on the scale of the balance.

♥ *Record*

1 Draw pictures to show your three ways of lifting. Describe each one, and make this chart to show your results.

Way of lifting	Reading on the spring balance
Straight up	grams
With one pulley	grams
With two pulleys	grams

2 What difference is there between using one pulley and two?

3 Did you have to pull further when you used two pulleys?

What pulleys are

Pulleys are wheels used with ropes or cables. They are useful on building sites. They let men pull downwards to lift things. This is easier than pulling upwards. Men can use their own weight to help them.

Lifts have a pulley at the top. A lift is connected by a cable to a heavy weight. It is called a *counterweight*. It weighs the same as the lift. When the lift goes up, the counterweight comes down. Its weight helps to pull the lift up. This saves engine power. When the lift goes down, the counterweight goes up.

You can see how this works by balancing weights on a pulley. It is easy to move them up and down.

Two pulleys make lifting easier. They cut the effort of lifting in half. But we have to pull the rope twice as far.

Picture 2 The more pulleys you use the less effort it takes to lift something

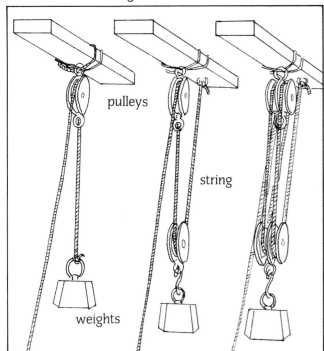

Picture 3 This crane has several pulleys. It can lift heavier loads than a crane with just one pulley

We can use still more pulleys. More pulleys make it even easier to lift something. But we have to pull the rope even further.

❤ To write

1 Describe how a pulley is shaped to stop the rope coming off.

2 Is it true to say that a fishing reel is a pulley?

3 You have a heavy weight to pull. You have pulleys and plenty of rope. There is a strong post fixed in the ground nearby. Say how you would move the weight, and draw a plan.

Making things easy with science

Wheels and friction

You used wheels, on page six, to cut down friction when you moved something. But even wheels are not perfect. A wheel turns on an *axle*. Where it touches the axle, there is friction. Scientists wanted to get rid of this friction, and make wheels spin more easily.

They invented *ball bearings*. Ball bearings are small steel balls which are packed round an axle inside the wheel. The bearings roll as the wheel turns. The wheel does not rub against the axle, but spins round easily.

Ball bearings are used in many machines. Bicycles and roller skates have bearings, to make them run smoothly.

We make bearings run even more easily by putting oil in them. This cuts down friction.

Where oil comes from

Oil is made by nature and lies deep underground among the rocks. To get oil, we drill down to these rocks. An area containing oil is called an oil field. Usually, the oil is at high pressure, and it spurts up the holes made by the drills. We put it into pipelines and ships, to take it to where we want it. If the oil does not have enough pressure to push it up the drill holes, we sometimes pump water or gas down to it. As the water or gas fills the oil field, its pressure pushes the oil up the drill holes.

How we transport oil

We carry oil around the world in huge ships called tankers. We take it to refineries, where it is turned into different kinds of things.

Some of it becomes lubricating oil, to

Picture 1 Machines do not work well if there is a lot of friction. Ball bearings help to cut down friction

Picture 2 The Shell/Esso North Cormorant Production Platform and a cross-section of an oil field.

go into machines to get rid of friction. Some is turned into petrol. Some is turned into diesel fuel, which is rather like petrol. Lorries and railway engines usually use diesel fuel. We also use oil to make plastic.

Oil pollution

If an oil tanker has an accident, oil is spilled into the sea. Because oil does not mix with water, but floats on the surface, it can do a lot of damage. It kills the living things on the surface of the sea. If the oil drifts to the shore, it ruins the beaches and kills the living things there.

The seabird in Picture 3 has been caught in oil. The oil makes its feathers cling together heavily, so that the bird cannot fly.

Picture 3 This guillemot is covered in spilt oil. It cannot fly or look for food

Part 2 Moving things with science

Sails

You will have seen the wind bending trees and driving the clouds. You have felt it blowing on your skin. People have used the wind to drive ships and machinery.

You will need a bowl of water, two matchboxes, cardboard, Plasticine, a cocktail stick with blunted ends, paper, scissors, elastic bands

♥ Experiment: How wind drives a sail

Roll paper to make two tubes. Use elastic bands to keep them rolled. You can blow through these tubes to make a wind. Now make two boats from matchboxes. They must be exactly alike. Fix a paper sail to one of them. Place the boats in the water, side by side. Blow through the tubes to race the boats. See which one moves better.

♥ Experiment: How wind can turn a spinner

Cut a disc out of cardboard 10 centimetres across. Cut it and bend it as Picture 2 shows. Pin it to a stick or sheet of cardboard. Make sure that it can spin easily. Blow on it, and see what happens.

Picture 2

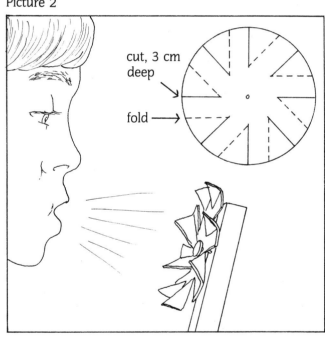

Picture 1 An experiment to find out how wind pushes a sail

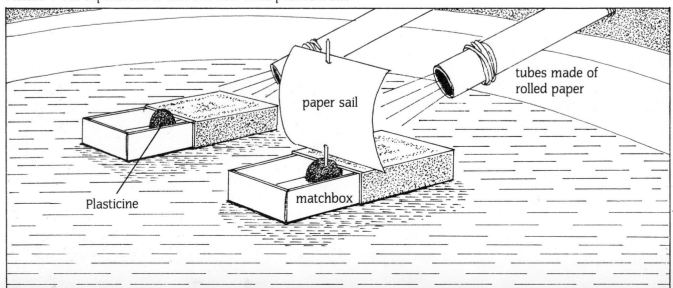

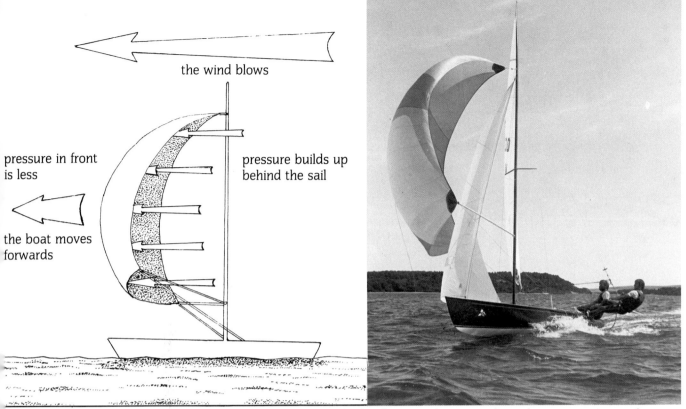

Picture 3 This yacht is being driven forward by the wind. The diagram shows you how the air pressure pushes at the sail and makes the boat move

♥ Record
1 Describe and draw each experiment.
2 Say what the wind is made of.
3 Does the wind make a sail move? Does the sail move faster when the wind blows more strongly?
4 Would it matter if a sail had holes in it? Why?

The wind
Air covers the Earth, and presses down on it. This pressing down is called *pressure*. The pressure is not exactly the same in every place. In some places the air gets warmer than in others. Warm air rises. This makes the air pressure less. We say that it makes an area of low pressure. The colder air nearby flows in to fill this area. This moving air is the wind.

Sails catch the wind. A sail gets in the way of moving air. Air builds up behind the sail. The air pressure behind the sail gets stronger than in front. The strong pressure tries to move into the weak pressure. As it does this, it pushes the sail. The sail moves forward.

Air pressure pushes at the sails of windmills. These sails are fixed to a spinner, and are twisted at an angle. They can move sideways, in a circle. The wind pushes them round and round. The sails are connected to machinery inside the mill. Some modern windmills are used to make electricity.

♥ To write
1 Do you think that the sails of a windmill want to move in a circle, or in another direction?
2 Can sailing ships go against the wind?
3 Wind blows from an area of _____ pressure to an area of _____ . (high, low, cold, warm)

Propellers

Propellers make ships and aeroplanes move. How do they work?

You will need a plastic propeller assembly, a bead, strong elastic, a ruler, drawing pins, two blocks of wood drilled through the middle, strong card, scissors, thin paper for streamers, Sellotape, pencils

♥ Experiment: To show that a propeller moves air

Arrange a propeller and elastic band as Picture 1(a) shows. Hold the apparatus still, and wind up the propeller with your finger. Then let go of the propeller. Hold some streamers near it and watch how they move. Put them in front of the propeller and behind it. They show how the air is moving. Try this experiment a few times. See if the air moves more strongly if you wind the propeller tightly.

♥ Experiment: To make a propeller pull itself along

Put pencils under the apparatus as rollers. (Picture 1(b) shows you how.) Wind the propeller up and then let go. The propeller will pull itself forwards.

If you have a toy boat with a motor, see how its propeller drives it forward.

♥ Record

1 Describe what you did and what happened. Draw a propeller, and the apparatus that you made.
2 Imagine that someone has never seen a propeller. Describe what one is like.
3 What does a propeller make air do? How does it do this?
4 Is the shape of a propeller important? Why are they shaped as they are?

Picture 1

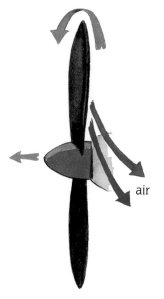

Picture 2 A propellor-powered aeroplane. The diagram shows how it works

How a propeller works

A propeller has sharp edges that cut through the air. Behind these edges, it is curved in a special way. This curve forces air backwards as the propeller spins. Your streamers showed how your propeller pushed air behind it.

As a propeller pushes air backwards, it moves forwards. A propeller claws its way forwards like someone swimming front crawl. As his arms move round, they push water backwards. This moves the swimmer forwards.

People got the idea of putting engines and propellers into gliders, to pull them along. The first people to manage this were Orville and Wilbur Wright. These American brothers made the first successful powered flight in 1903.

The first ships driven by engines had huge paddle wheels on their sides or at the back. The engines turned the paddle wheels and they drove the ship forward. Later, people put propellers on ships because propellers worked better. In 1845, a ship called the *Great Britain* became the first propeller-driven passenger ship to cross the Atlantic.

♥ To write

1 Why are a ship's propellers put at the bottom of its hull?
2 What does a propeller need if it is to work properly?
3 What else can a propeller be used for?

Aeroplanes and parachutes

Aeroplanes let us fly through the air. Parachutes let us fall safely. How do they do this?

You will need Plasticine, string, a handkerchief, a larger square of cloth, scissors

Picture 1 Your experimental parachute should look like this

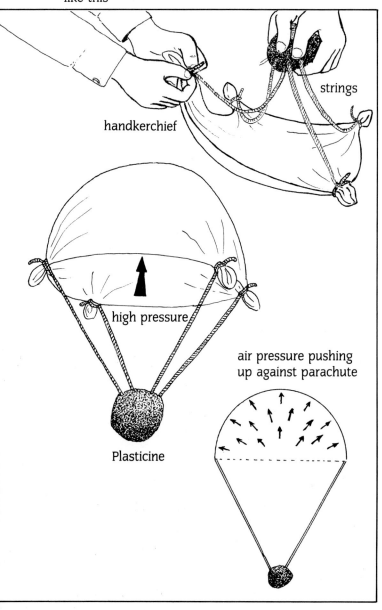

♥ Experiment: How a parachute works

Experiment with a falling object. Take a lump of Plasticine as big as your fist. Find a clear space, and toss the Plasticine into the air. Count how many seconds it takes to fall. Now take a square of cloth the size of a large handkerchief. Tie 30 centimetres of string to each corner. Fix the loose ends into the Plasticine. Toss the Plasticine into the air again, as high as last time. Count how many seconds it takes to fall.

Take a much larger square of cloth. Use that as a parachute. See if the Plasticine falls at the same speed. Try the experiment a few times.

♥ Record

1 Make this chart to show how fast the Plasticine falls.

	Time taken to fall
The Plasticine on its own	seconds
Plasticine with small parachute	seconds
Plasticine with large parachute	seconds

2 Draw your experiment and describe what you did.
3 Say what effect a parachute has on how something falls.
4 Would a parachute work if there were no air?
5 What difference would it make if the parachute were full of holes. Why?
6 Why does a big parachute work better than a small one?

Picture 2 The air pressure pushing up inside the parachute slows it down so the person reaches earth safely

Air pressure

If you blow up a balloon, it swells and may explode. Your blowing builds up strong pressure inside the balloon. This strong pressure pushes outwards towards the lower pressure outside. This push can be strong enough to burst the balloon. Strong pressure always pushes towards weak pressure.

A parachute is shaped to trap air as it falls. This air gets squashed up, or *compressed*, inside the parachute. The compressed air builds up a strong pressure. The strong pressure inside pushes towards the weaker pressure outside. It pushes up at the underneath of the parachute. This upward push slows down the fall.

Aeroplane wings

The wings of aeroplanes are carefully shaped. The top of the wing lets air flow easily over it and away from it. This causes a low pressure of air above the wing. But the bottom is tilted and curved, to slow down the flow of air. This builds up strong pressure under the wing. This strong pressure underneath pushes towards the weaker pressure above. It lifts the aeroplane upwards.

♥ To write

1 Describe a common use of compressed air.
2 What can you say about air which is suddenly slowed down?
3 Say how a bird uses air pressure.

Picture 3 Aeroplanes have specially-shaped wings so that the air pressure lifts them up

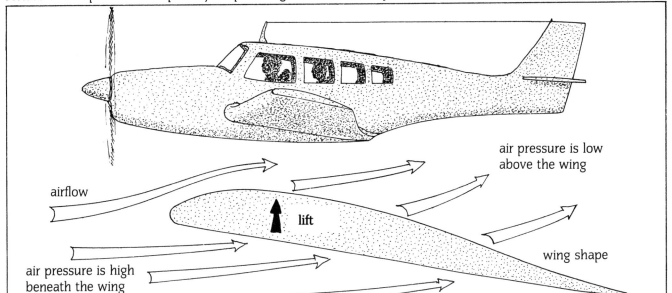

Steam power

Steam does not look strong, yet it can drive huge machines.

> You will need a metal tin with a tight lid and a small hole in its side, a wire cradle to hold it, candles and matches, card, wire, a plastic straw, plastic tubing

Picture 1 Arrange the experiment like this. **Take great care as steam can burn you**

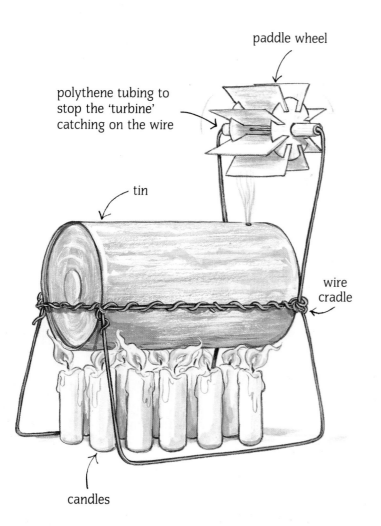

♥ Experiment: To show steam at work

This experiment should be done out of doors, in a clear space. An adult **must** be with you.

Take the metal tin with a hole in it. Check that the hole is clear. Put two or three centimetres of water into the tin and put its lid on firmly. Stand the tin over candles in a wire frame, as the picture shows.

Make a paddle wheel from cardboard and a straw. Use wire to hold the paddle wheel over the hole in the tin.

Light the candles and wait. Steam will come from the hole. After a while, it will spin the paddle wheel. Do not stand close to the experiment. An adult must put out the candles and clear away afterwards.

♥ Record

1 Describe and draw the experiment. Say what happened.
2 How was the steam made? What do you always need to make steam?
3 What would happen if the tin had no hole?
4 What made the paddle wheel turn? How can we use this idea in machines?

Picture 2 Robert Stephenson's 'Rocket', the first steam engine. The diagram shows how the engine worked

Steam pressure

When water gets very hot, it boils. It bubbles and starts to turn into steam. Steam takes up far more space than water. Water expands when it turns into steam. Normally, the steam just goes into the air. But if we boil water in a sealed container, the steam cannot get away. A high pressure of steam builds up. If it cannot escape, it can burst the container.

Steam engines use steam pressure to move things, like the jet of steam drove the paddle wheel. They shoot steam into special *cylinders*. Each cylinder has a piece of metal inside called a *piston*. Steam pressure makes pistons move up and down inside cylinders.

We connect the pistons to machines, to make them work. In the picture of an early railway engine, you can see the cylinders and pistons.

Steam turbines

Power stations use steam to spin large wheels in machines called *turbines*. The wheels have curved blades, like propellers. The spinning turbines drive *generators* to make electricity. Power stations boil huge amounts of water to make steam, but they do not waste it. They let the steam cool until it turns back into water. Then they heat it again to make fresh steam.

❤ To write

1 Water turns into steam when it _____. (expands, heats, boils, bursts, cools)
2 Where at home can we see steam pressure at work?
3 What were steam engines once used for?

Electric power

How does electricity make a motor work?

> You will need two strong magnets, two wires, a strong battery, the coils and shaft of a toy electric motor, plastic straws, Sellotape, books to support the magnets and motor

❤ Experiment: Making electricity move a wire

Set up two magnets so that the north pole of one is opposite the south pole of the other. There will be a magnetic field between these two magnets. Take the coils and shaft of a toy electric motor. Notice that each coil is connected to a curved piece of copper on the shaft. Use plastic straws and Sellotape to hold the shaft and coils between the magnets. Connect two wires to a strong battery. The wires must have bare ends, of course. Touch the ends of the wires to the copper pieces on the shaft. The coils and shaft will spin. Take away the magnets and try the experiment again. See if the coils will move without the magnets.

❤ Record

1 Describe your experiment. Draw what you did. Describe what happened.
2 Is there a link between electricity and magnetism?
3 What effect did electric current have on the wire coils?
4 Can an electric current in a magnetic field produce movement?
5 Might this work the other way round? If you moved a wire in a magnetic field, might you produce electricity in the wire?

Electricity

Electricity and magnetism are closely connected forces. Electric motors and dynamos work by using this connection.

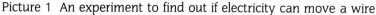
Picture 1 An experiment to find out if electricity can move a wire

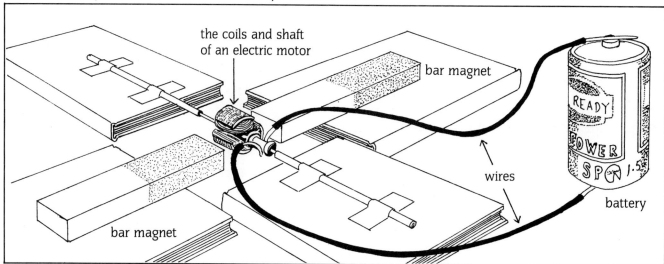

Study the electric motor in Picture 2. It has wires leading into it, to carry electric current. Inside the motor, the wires are connected to coils. The coils are made of loops of wire that are fixed to a shaft. The coils and shaft can spin.

Strong magnets surround the coils. This means that they are in a strong magnetic field. When the motor is switched on, electricity flows through the coil. The result is that the coils and shaft start moving round. They keep spinning until the electricity is turned off. The turning shaft can be fixed to anything: a wheel, or a drill.

A bicycle dynamo

A *dynamo* is built in much the same way. There are magnets surrounding coils of wire. The coils of wire are on a shaft that spins. They spin when the shaft turns. As they spin in the magnetic field, electricity is made in the wire. Other wires take this electricity to the front and rear lamps.

Both motors and dynamos sometimes have the magnets on the shaft and the wire coils on the outside. They work just as well when they are arranged like this.

♥ To write

1 Electricity and magnetism are _____ . (opposite, equal, connected)
2 An electric motor spins _____ inside _____ . (coils, magnets)
3 What things at home use electric motors?

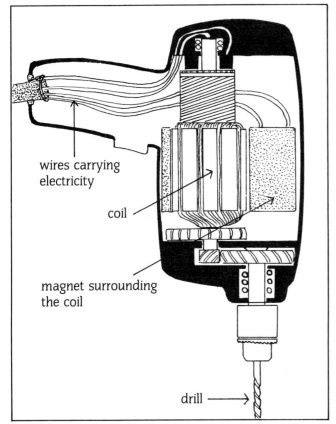

Picture 2 An electric motor drives this machine

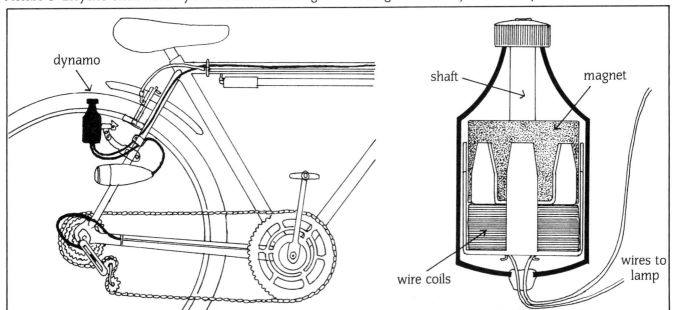

Picture 3 Bicycles often have dynamos to work the lights. The diagram shows you how a dynamo works

Jet and rocket engines

Picture 1 A 'jet' balloon

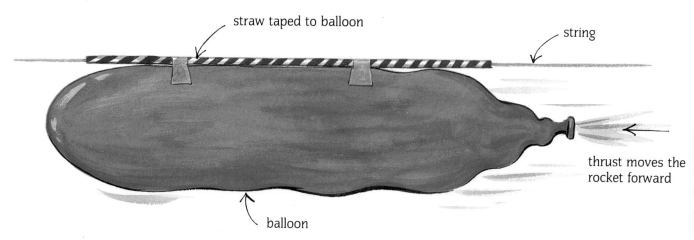

How do jet and rocket engines move things?

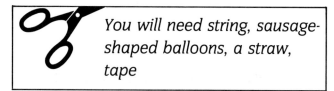 *You will need string, sausage-shaped balloons, a straw, tape*

♥ Experiment: How to make a jet of air

Slide a straw onto a long length of string. Tie the ends of the string to two places far apart. Make the string tight, so that the straw can move along it easily. Next, blow up a sausage-shaped balloon and holds its neck to keep in the air. Tape the balloon to the straw, and then let go of it. See how far and fast it travels.

Repeat the experiment several times. Try blowing less air into the balloon. See if this affects its speed and distance.

♥ Record

1 Describe and draw your experiment.

2 Make this chart:

Amount of air in balloon	How far it went
As much as possible	metres
Quite a lot	metres
Not much	metres

3 What made your balloon move forwards?
4 Why does the amount of air inside the balloon matter?
5 Would the balloon move if there were no air around it?

Action and Reaction

A simple action

Every action causes a *reaction*. Imagine yourself standing on roller skates carrying a heavy bag. If you throw the bag away from you, you will roll backwards. You will move in the opposite direction from the bag. You make an action when you throw

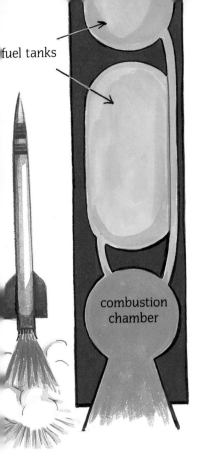

fuel tanks

combustion chamber

pressure is released through the exhaust nozzle

Picture 2 This is how a rocket engine works and pushes the rocket forward

the bag. The *reaction* makes you roll the other way. This is an important rule in science. We say that *every action causes an equal and opposite reaction.*

Reaction from a jet

When you let go of your full balloon, a jet of air rushed out. This action of the air made the balloon move the other way. It caused a reaction by the balloon.

If there had been no air in the room, the balloon would have gone even faster. (If it hadn't exploded!) The air inside it would have rushed out faster and made a stronger reaction.

Jet and rocket engines

Jet and rocket engines work because of this rule. They send out hot gas at one end, like your balloon sent out air. This action causes a reaction. The engines move in the opposite direction. As the gas shoots out backwards, the jets and rockets move forwards. The main difference between your balloon and these engines is that engines are millions of times more powerful.

There is more about these engines on page 33.

♥ To write

Choose the best words to complete these sentences:

1 Every action causes a _____ . (fall, change, reaction, explosion)
2 A reaction is always _____ to the action. (similar, opposite, different)
3 What makes jet engines move forwards?

Picture 3 A rocket launch

Moving things with science

Modern engines let us move heavy things a long way. Here are three types of modern engine.

The petrol engine

Just as water can turn into water vapour, petrol can turn into petrol vapour. This vapour can be made to burn. When this happens, it expands tremendously. It can make a huge pressure. This makes it ideal for an engine.

Inside a petrol engine are hollow cylinders. Each cylinder has a piston in it that can move up and down. Each piston is connected to a *crankshaft*. The crankshaft is connected to other pieces. In a car, these drive the wheels round.

Petrol vapour is drawn into the cylinders. Air is mixed with it, because things will not burn without air. A spark is made in each cylinder in turn. This explodes the mixture and makes a huge pressure of gas. The pressure slams each piston down its cylinder in turn. Each piston moves the crankshaft round a bit. As each piston goes down in turn, it turns the crankshaft a bit further. As the crankshaft keeps on turning, it pushes the first piston back to its starting place.

Picture 1 This diagram of a petrol engine shows how it works

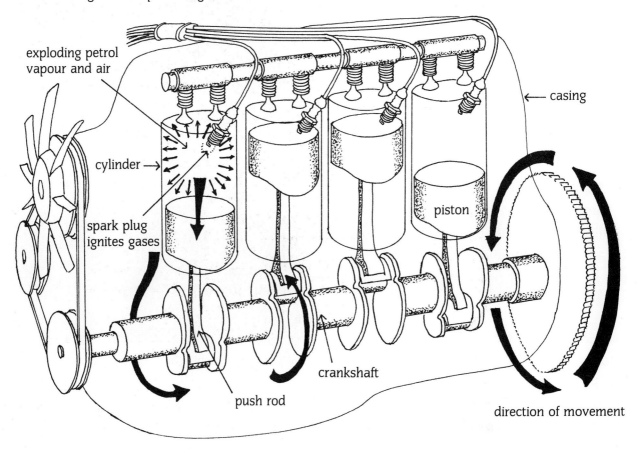

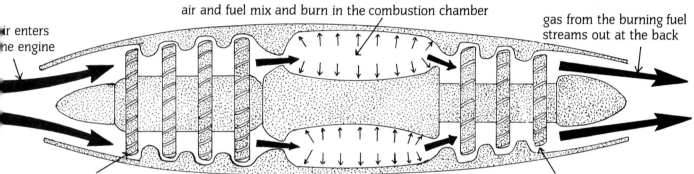

Picture 2 This aeroplane has jet engines. The drawing shows how the engine works and pushes the aeroplane forward

The jet engine

A jet engine uses a fuel called kerosene, which is like paraffin. It burns this fuel with air in a *combustion* chamber. ('Combustion' means 'burning'.) The fuel expands powerfully in the combustion chamber and builds up a huge pressure.

There is an opening at the back of the chamber. A jet of hot gas rushes out through it. This release of gas causes a forward force called *thrust* in the engine. The thrust makes the engine move forwards. A jet engine carries fuel with it, but takes in air from the atmosphere. An aeroplane pilot changes the power of the thrust by changing the amount of fuel and air going into the combustion chamber.

The rocket engine

A rocket engine has to work in airless space. It takes its own 'air' with it. Scientists make oxygen so cold that it turns into liquid. In this way, masses of oxygen can be stored in a small tank.

In a rocket engine, rocket fuel and liquid oxygen are sprayed into a combustion chamber together. They burn fiercely and expand to make an enormous pressure. Many early rockets blew up because the pressure broke the combustion chamber. The gas escapes through a nozzle, and causes thrust, which drives the rocket forward.

Part 3 A scientific look at nature

A visit to a pond

Ponds are full of life.

> You will need some containers, a reference book, a magnifying glass, a notebook, a rake, a mud net with a firm wire scoop, a pond net with a soft mesh scoop.

How to explore a pond

You must not go to the pond without an adult.

Look at the weeds growing round the edge of the pond. Use the rake to pull weeds from under the water, but do not pull up more than one or two. Use your reference book to try and identify them. See if any creatures are clinging to the weeds.

Drag your pond net across the surface of the water to catch any creatures there. Tip them out carefully into a container with pond water in it. Study them through your magnifying glass. Use your reference book to identify them, and record what you find.

Then scoop lower down, into the middle of the water. Catch any creatures that are floating in the water.

Lastly, use your mud net to scoop the bottom. Catch the creatures that live on or in the mud and rotting plants.

Remember to put all your samples back in the pond when you have finished.

♥ Record

1 Describe your expedition to a pond, and what you found.
2 Make a large wall picture of a pond, as

Picture 1

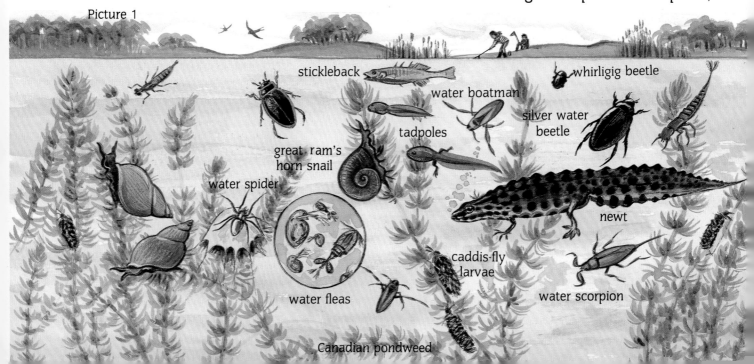

if cut down the middle. Collect pictures of different kinds of pond life, and stick them on the picture in the right places.

Plants in a pond

Most ponds have many different plants in them. This book shows only a few.

Some, such as reeds and rushes, grow round the edge of the pond. The roots of these plants keep the banks of the pond firm.

Some plants float on the water. Some plants seem to be floating, but have long stems reaching down to roots in the mud at the bottom of the pond. Other plants grow under the water.

Plants are vital to life in the pond. They make hiding places for many of the animals, and they make oxygen for the pond creatures to use.

Pond Animals

Many pond animals eat water plants like the tiny *algae*. Such animals are called *herbivores*. Some animals eat dead plants and dead animal bodies. These are

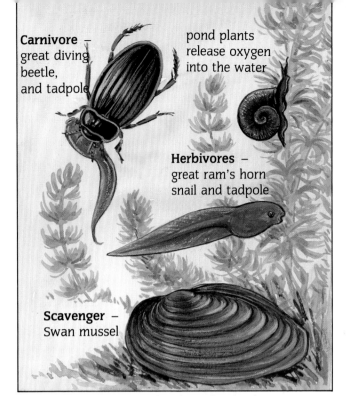

Picture 2 The 'food web' of pond animals

scavengers. Some live by eating other animals. These are *carnivores*.

♥ To write

1 How do plants help keep the pond's edges firm?
2 Is it true to say that scavengers are important animals? Why?
3 Name some herbivores and carnivores, and say what they eat.

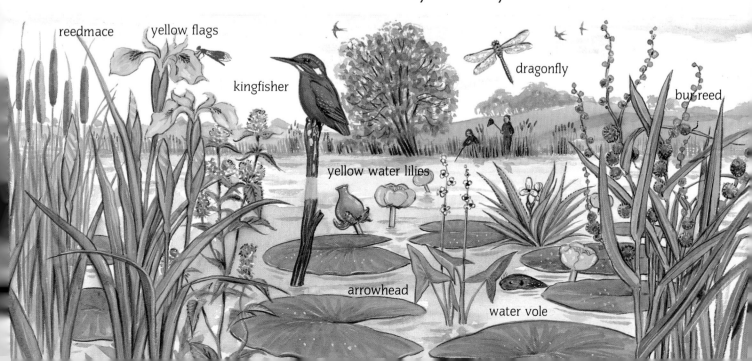

Make an aquarium

You can keep water animals in the classroom for a while, before returning them to a pond.

 You will need one or two glass-sided tanks, gravel, water plants, pond animals, frogspawn

♥ Keeping animals in an aquarium

Put washed gravel 5 centimetres deep in a glass tank. Put water plants into the gravel. (You can buy them.) Pour water gently into the tank. Leave the tank for a few days, and then put in your animals. Do not put in any carnivores.

♥ Raising frogs from frogspawn

Put frogspawn into a separate aquarium. Watch every day as the little black eggs change shape and later wriggle out of their jelly. Later, they will start to feed on the algae that live on the water plants. Watch their bodies develop. They will need something in the water to jump from. But once they are jumping properly, it is time to let them go. They need to catch lots of tiny insects in order to live.

♥ Record

1 Draw your aquarium and the plants and animals that you put in it.
2 Say how the animals live. What do they eat? Do they come to the surface to

Picture 1 The 'life cycle' of a frog

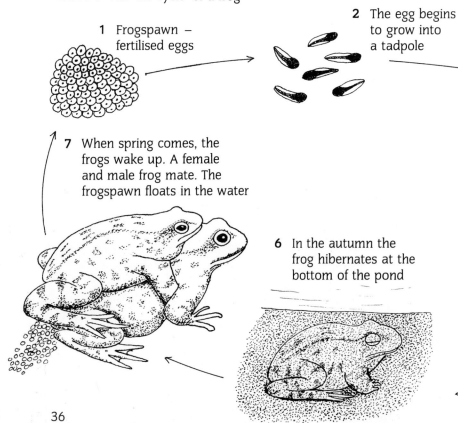

1 Frogspawn – fertilised eggs
2 The egg begins to grow into a tadpole
3 After a few weeks the tadpole grows back legs
4 Soon small front legs grow and the tail begins to shrink
5 The frog develops lungs. It can now breathe on land and live out of water
6 In the autumn the frog hibernates at the bottom of the pond
7 When spring comes, the frogs wake up. A female and male frog mate. The frogspawn floats in the water

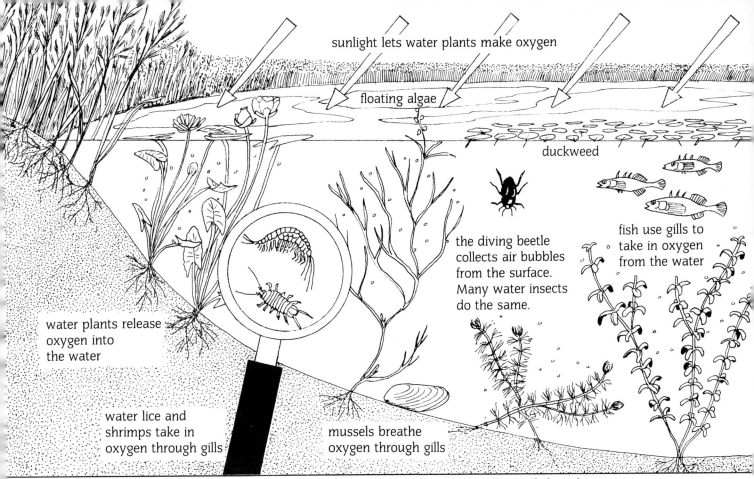

Picture 2 Animals need oxygen to live. Pond plants give out oxygen and the pond animals breathe it in

breathe? How do they move? Do they grow much?

3 Describe how frogs develop. Draw them whenever you notice a new change.

Water life

All animals need oxygen to live. Most water animals have *gills* which take oxygen out of water. Most of the oxygen in water is made by plants in the water. Water plants have chlorophyll, just like land plants. Chlorophyll lets plants make oxygen. Even tiny plants like algae can do this.

A water animal begins life as an egg. The egg is surrounded by a jelly, which is food for the animal as it starts to grow. After a time, the animal breaks free of the jelly. It starts to eat algae and develops gills to 'breathe' oxygen from water. It grows slowly to its full size. Later, male and female animals will make new animals. Females make eggs and males make sperm. Sperm from a male must join an egg from a female to make a new animal. Water animals usually lay eggs and sperm in the water, where the joining, or *fertilisation*, happens. Once they have been fertilised, the eggs start to grow into new animals.

♥ To write

1 Find four pairs of words: female, sperm, gills, eggs, algae, male, oxygen, plant.
2 Where does the oxygen in pond water come from? What happens if there is none or if a lot of it is used up?
3 Tell the story of the 'life cycle' of a water animal.

A visit to a seashore

Lots of things live in the sea and beside it. How many things can you find on the seashore?

> *You will need a notebook, a reference book, a magnifying glass, a jar and cloth, collecting bags*

♥ How to explore a seashore

Look at the shore. Can you see a high-tide mark? Is the tide in or out? See what the beach is like, and how smooth the pebbles are. See what shells you can find. Fill a jar with sea water. See how cloudy it is. Strain the water through a cloth and examine what you find.

Look for plants. Some will be land plants, high up the beach. Lower down, there will be seaweeds. See if they are all alike, or if there are differences. Take some samples. Put a piece of seaweed into a jar of seawater. See if any animals come to life.

Look for animals. Look among seaweeds. Look in shells. Look in pools and under rocks. Scoop sand away with a stick from any holes you see.

♥ Record

1 Describe the seashore, and pin up a map of it.
2 Label your samples, and make a display of them.

Seashore life

Sea water is usually cloudy. It has many things in it, apart from salt. It has many plants and animals that are too small for us to see on their own, but which show up as a cloud.

Above high-tide level, land plants hold together the sand and stones with their roots. Seaweeds grow further down the beach. High up, seaweeds are green. Lower down, they are brown. Further down still, they are red-coloured. The reason is that the seaweeds furthest up get quite a lot of sunlight. They have green *chlorophyll* like land plants. Seaweeds further down the shore spend most of their time in deep water where the light is dim. They have other things in them as well as chlorophyll. These things help the seaweeds to catch this dim light better. They give the seaweeds their different colours.

Sea animals can be scavengers, herbivores, carnivores or *omnivores* (which eat anything).

Sea shells are the remains of animal life. They come in two types: the snail-type shell and the flatter *bivalve* type. The picture shows how these shells look when they have living animals inside them.

♥ To write

1 How could you have a close look at the tiny plants and animals in sea water?
2 What do you find on a seashore as well as plants, animals, stones and sand? Where does it come from?

Picture 1 The seaside. Can you see the water line in the bottom picture of the rock pool?

A wood in winter

Deciduous trees lose their leaves in winter
Coniferous trees keep their leaves all year round

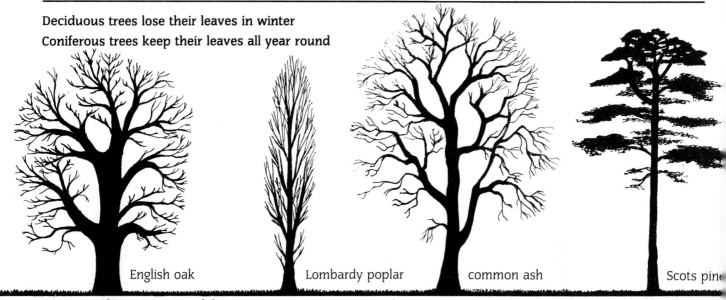

Picture 1 These are some of the trees you may see in a wood

You will need a notebook and pencil, a camera, bags for carrying samples, card and plaster of Paris for making casts, a trowel, and a reference book

Are woodlands completely dead in winter? Go and find out. NB You must go with an adult. Stay near a landmark, like a stream or large rock, so that you do not get lost.

Look for some plants which have leaves and some which do not have leaves. Use a reference book to identify them. Collect sample dead leaves and twigs. You can make plaster casts of them. (Plaster casts are described in Book 2.)

Identify any plants that are still green. You can take samples of mosses.

Look for traces of animals – nests, holes and tracks. You may be able to make plaster casts of tracks. Put card around the imprint, pour in plaster and leave it to set. Then dig up the whole thing, including the soil. When it has grown really hard, you can brush off the soil.

Dig the ground and see last year's leaves rotting into soil. See what 'soil animals' are in it.

♥ Record

Make a display about your visit. Pin up a map of the area. Describe the woodland,

Picture 2 The tracks (footprints) of different animals

badger

common shrew

red deer

ferret

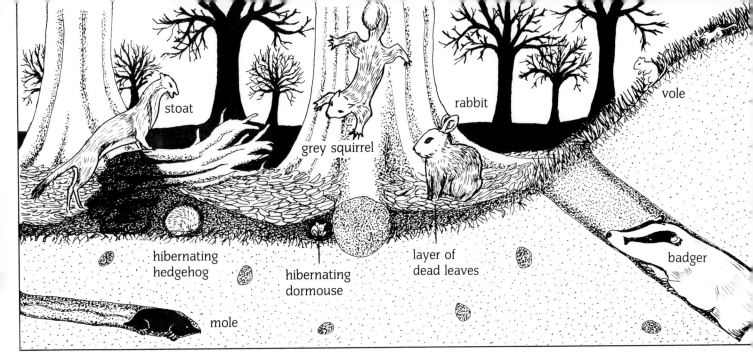

Picture 3 Look out for these animals when you go to a wood

and how you explored it. Put up photographs and drawings that you brought back, explaining what they show. Put out any samples you have of fur, bones, pressed leaves, and so on.

Plants

Trees which lose their leaves each winter are called *deciduous*. The ones which keep their leaves are called *evergreens*.

Look at the twigs of the deciduous trees. You can see the buds for next year's leaves. The *girdle-scars* show where the twig ended last year. You can see how much the twig has grown in a year.

Few large plants remain green in winter. Grasses and mosses do, but they stop growing. Almost all woodland flowers disappear, but do not die. Their *bulbs* stay alive underground each winter. These plants, which come up year after year, are called *perennials*.

Animals

Some animals stay active during the winter. Others go into a deep sleep called *hibernation*. This sleep is almost like death – it slows the animals' hearts and breathing, so that they need hardly any food to stay alive. They use up their stores of body fat as food. They may wake up and come out on warm days.

Many birds *migrate* to warmer countries in winter.

♥ To write

1 Make lists of the names of deciduous and evergreen trees.
2 Find three groups of words: hedgehog, bluebell, hibernate, squirrel, swallow, migrate, primrose, swift, perennial.

fox

weasel

hedgehog

grey squirrel

A wood in summer

You will need collecting bags, a notebook and pencil, a camera, a ruler, a reference book

Are woods just collections of trees? Or are they more than that? Study a wood to find out.

Pick out a landmark and stay near it.

Look at the plant life. See how many different types of plant are growing in your area. Include all the plants, from the trees down to the tiny mosses. Use reference books to identify each type of plant.

You can take ferns, mosses and leaves from trees as samples, but do not pick flowers or berries. Where possible photograph plants and interesting objects such as fallen trees. If you cannot take photographs, make drawings instead.

Look for animal life, from soil creatures in the ground to the birds above. Look in the grasses, under rocks and up among the branches. Look for clues on the ground, such as animal droppings or feathers, fur and bones, or gnawed nuts and bark. Look for holes in the ground. Measure them, and think what animals might live inside.

Picture 1 A wood in summer

♥ Record

1 Back in school, press any leaves. Label them, and mount them beside drawings of the plants they came from.
2 Make a display of photographs that you took. Label them with the names of the plants that they show.
3 Make a table display of any gnawed nuts, feathers, mosses and so on that you found.
4 Pin up a map showing the area you explored, and a description of the area.
5 List the different animals and plants that you found. Display the list with pictures and facts to go with it.

Woodland life

The picture shows you some of the most common types of woodland plants. It also shows you some of the animals that were round you while you explored. You probably saw only a few, because animals are good at hiding.

The animals and plants that you saw are connected in an important way. They need each other for food and shelter.

Woodlands and forests are important for people. There are thousands of visitors every year, and special arrangements have to be made for them.

♥ To write

1 A fox is a _____ . (carnivore, herbivore, omnivore)
2 A badger eats _____ and _____ as food.
3 In what ways do people damage woodland life? How can we keep the damage down?

A study of wild birds

You need to hide to watch wild birds where they live. It is easier to get them to come to you.

> You will need suitable pieces of wood, hammer and nails, saws and drill, kitchen scraps, wild bird seed, a small jar for water

Picture 1 You will need these pieces to make the bird table

♥ Build a bird table

Make a table like the one in Picture 1. (Ask an adult to help you.) Leave gaps in the rim to let rainwater drain away. Put the table high up, where cats cannot reach it.

Put kitchen scraps on the table, and also some seed. You can buy this from pet shops. Make sure that there is always fresh water in the drinking jar.

Watch your bird table at various times, and observe the birds that come to feed. Use a reference book to see what kind they are.

♥ Build a nesting box

Try to get some birds to live near you. Make nesting boxes to encourage them. Make two or three boxes like this for the school grounds or your garden at home. Give each box a different-sized opening. Put the boxes as far apart as possible. See if any birds make their nests in them. You can watch them raising their families.

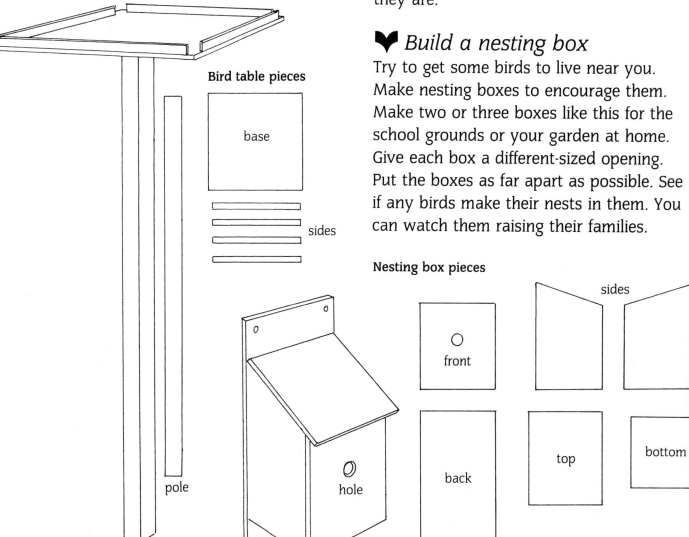

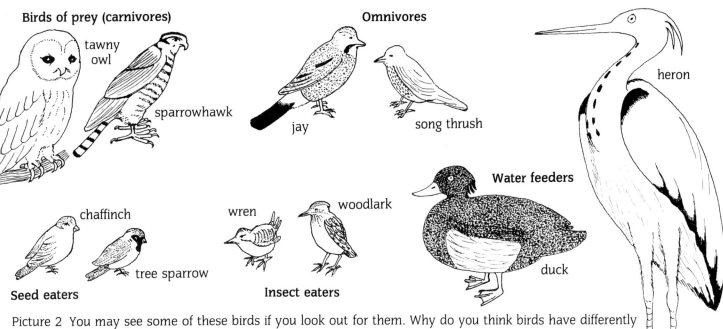

Picture 2 You may see some of these birds if you look out for them. Why do you think birds have differently shaped beaks?

Day	Time	No. of birds at the table	Types	How they behaved	Interesting observations
Tuesday 13th	9am	3	2 sparrows 1 starling	The sparrows were frightened because the starling bullied them.	One of the sparrows had hurt its wing.

Chart 1

 Record

1 Draw and describe the things that you have made.
2 Make Chart 1 to record your observations.

Types of birds

Carnivores, or birds of prey, eat living animals, and will not come to your table. Owls, hawks, falcons and eagles are carnivores. *Scavengers* such as crows might come, but they usually eat dead bodies, which we call *carrion*. You are unlikely to see water birds, although gulls will come if you live near the sea.

Birds that eat parts of plants, such as berries and seeds, will come. So will those that eat many different kinds of food. So you may get sparrows, finches, starlings, tits, robins, and thrushes. Use reference books to learn the different types.

Some birds will not appear in winter because they migrate. They fly away to warmer places for the winter. Swallows and swifts migrate.

Some birds appear only in winter and then they fly away again in spring, like geese.

There are about 8600 different types of bird in the world. You should be able to spot several dozen in a year.

To write

1 Owls, hawks and falcons are all _____. (scavengers, carrion, carnivores, omnivores)
2 Why do some people say that wild birds should not be fed, except in very bad winter weather?
3 Some birds come to Britain for the winter. Why do they do this?

A scientific look at nature

Earth, moon and sun

The Earth is a large ball travelling through space. It is a planet. It travels round the sun with eight other planets. The Earth is about 150 million kilometres from the sun. It takes light from the sun over eight minutes to reach us. This means that if the sun suddenly stopped shining, we would see its light for eight minutes afterwards.

The Earth has a moon circling round it. The moon is a ball of rock, much smaller than Earth. It has its own *gravity*, which is much less than that of Earth.

The effect of the sun

All life on Earth depends on the sun, because sunlight makes plants grow. All green plants have chlorophyll in them. Chlorophyll lets a plant use air, water and sunlight to make *sugar*. Sugar is the food that makes plants grow. This way of using sunlight to make food is called *photosynthesis*.

Photosynthesis makes oxygen as well, and uses up carbon dioxide from the air. Animals need to breathe oxygen in order to live. This means that all animals on Earth, including us, depend on plants for fresh air as well as food.

The effect of the moon

The gravity of the moon affects the seas that cover most of the Earth. The moon pulls at the water on Earth. It pulls the water upwards very slightly when it is overhead. This raises the level of the sea. As the moon travels past on its journey round the Earth, the water sinks down again. It is this change that makes high tide and low tide happen. The sun pulls at the oceans too, but not as strongly as the moon does.

Picture 1 The pull of the moon affects the seas on Earth. It makes the sea move up and down. We call the movements high tide and low tide.

Picture 2 All life on Earth depends on the sun

The stars

The stars are too far away to affect us. It takes light from the nearest star over four years to reach us. They show us how big the world of nature is. There are many more stars than you can see.

Scientists know that every star is a sun, rather like our own. Some may have planets with life on them. At the moment, we can only guess. The stars are too far away for us to know their secrets.

New words

algae   tiny water plants that have no roots, stem or leaves
axle   a rod connecting wheels

ball bearing   balls put in the moving parts of machines to cut friction
bivalve   a sea creature which lives in a hinged shell
bulb   a part of a plant that lies underground from year to year

carnivore   a meat-eating animal
carrion   dead flesh
chlorophyll   the green stuff in plants that lets them make starch from sunlight, air and water
combustion   burning
compressed   squashed or squeezed
counterweight   a weight used with a pulley to help lift something
crankshaft   a specially-shaped piece of metal to change the up-and-down movement of pistons to round-and-round movement
cylinder   a tube in an engine that holds a piston

deciduous   losing its leaves in winter
dynamo   a machine which produces electricity from movement

expands   grows bigger

friction   a force that causes drag when two things rub against each other
fulcrum   the anchor that a lever rests against, or on

gears   wheels of different sizes that make work easier
generator   a machine which produces electricity from movement

gills   the parts of water animals that take oxygen from water
gravity   the force that pulls things towards a large object in space

herbivore   an animal which gets all its food from plants
hibernation   the deep sleep which takes some animals right through winter

lever   something used to move an object

migrate   to travel to a different land each year

omnivore   an animal that feeds on many kinds of food

perennial   living year after year
photosynthesis   the process by which plants use sunlight to make starch and oxygen
piston   a part of an engine that moves up and down inside a cylinder
pressure   a pushing force
pulley   a wheel used with rope to help move things

reaction   a force caused by an action

scavenger   an animal which eats dead and decaying stuff, or carrion
sperm   stuff that a male animal puts with a female's eggs, to start them growing
sugar and starch   the food that green plants make by photosynthesis

thrust   the force which moves something
turbine   a paddle wheel turned by power such as steam or water to drive a generator